ICELAND:
THE PUFFIN EXPLORERS
BOOK OF FUN FACTS

RA ANDERSON

Iceland: The Puffin Explorers: Book of Fun Facts

ra-anderson.com
myfavoritebookspublishingco@gmail.com
My Favorite Books Publishing Company, LLC.
Kingston, Georgia USA

Ordering Information:
Quantity sales. Special discounts are available on quantity purchases by corporations, associations, and others. Orders by U.S. trade bookstores and wholesalers. For details, contact the publisher at the address above.

Proofreading by The Pro Book Editor
Interior and Cover Design by IAPS.rocks
Photography by RuthAnne Anderson
Contributed photos by Alan Rohrbach
Sketches by Hannah Jones

ISBN: 978-1-950590-13-1

Main category—JUVENILE NONFICTION/Animals
Other category—JUVENILE NONFICTION/Science & Nature

First Edition

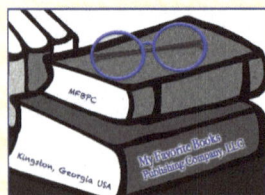

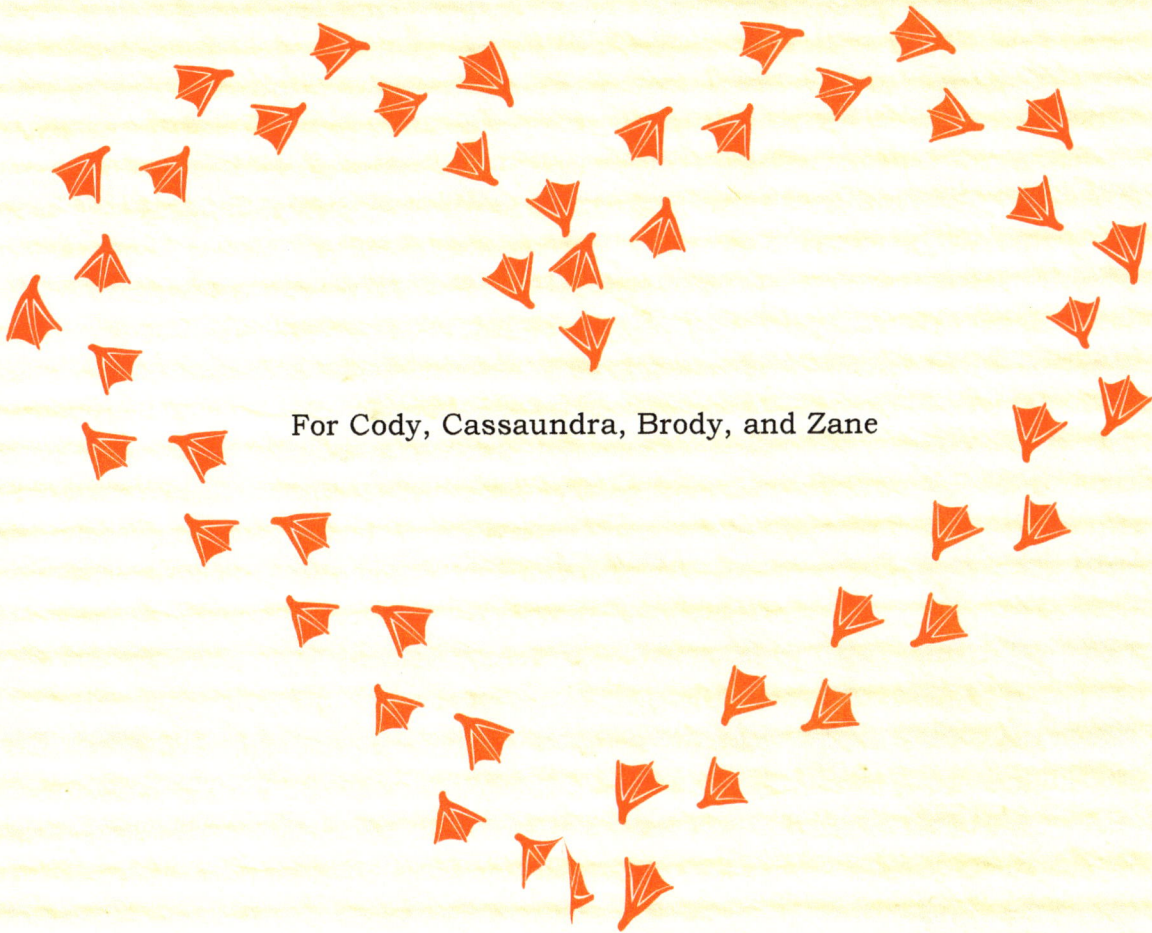

For Cody, Cassaundra, Brody, and Zane

INTRODUCTION

Dear Readers,

When traveling to Iceland for the first time, I didn't know I would find my heartstrings being tugged by a one-pound bird! In flight, a puffin looks like a black-and-white flying football, and when landing, it looks like a fumbled, tumbling ball. Its comical stance and little marching walk...its triangular-shaped eyes looking up, around, and back at me for another glance while cooing at its one true love... A puffin seabird simply stole my heart!

In 2018, I spent a total of four weeks in Iceland, preparing for and then attending my eldest son's Icelandic destination wedding. During these two trips, I completely fell in love with the country and my new puffin friends.

In 2019, I returned to Iceland for another couple of weeks to capture even more pictures of this extraordinary country. Iceland is such a beautiful place that I couldn't keep these images all to myself. Since I had gathered so many facts about Iceland and there wasn't enough room for all of them in Iceland: The Puffin Explorers Series, I decided to write a fourth book only about the fun facts just to share them with you.

However, there were changes from my time spent in Iceland in 2018 to my 2019 visit. Glacier Lake goes back farther toward the mountains. The icebergs are plentiful but not the huge, towering beasts they were the prior year. Fewer puffins were spotted (fewer by hundreds) and it was sunny and warm the whole time we were there. These are simple observations from a non-scientist's view of how fast even a place named for ice is melting away. No matter if you believe what scientists are saying or don't believe climate change is real, it is happening in front of your eyes. Greenland is a prime example with its ice sheet losing 248 billion tons of melt during the first seven months of 2019, causing sea levels to rise. Glaciers are melting until they become so thin that they cannot move on their own, which is a requirement to qualify as a glacier. Consider the melting Icelandic glacier Okjökull, which lost its glacier status (*jökull* is Icelandic for "glacier,") and is now simply Ok—but it's not okay. These melt-offs are causing unknown amounts of damage to the Arctic ecosystem and the rest of our planet. All I hope is for people to see what is really happening, to be open to ideas for solutions, and to care for the world. As the saying goes, there is no Planet B!

Plastic! I am guilty for using plastic myself. There are plastic items I use every day, like plastic sandwich bags, water bottles, Toothpaste tubes, my cat treat containers, milk jugs, and even some clothing material has plastic. What I didn't know is that even recycled plastic is making its way into our ecosystem and doing great harm. Every day on social media we see photographs or videos of sea critters, birds, fish, and whales that were harmed or killed by our plastic. Today, I challenge you to simply make yourself aware, a mental note, of the everyday plastic you are using. My second challenge is that you not only recycle and reuse but try and purchase alternative items. The simplest changes can make a big difference, like switching to reusable grocery bags. And, one that most wouldn't think of as being an issue is drinking straws—do you really need them and if so, why not use eco-friendly straws that more and more restaurants are switching to as well? I'm not asking you to change the world and definitely don't stop brushing your teeth with toothpaste, but maybe make it safer for our critters...one straw at a time.

I hope you find these bits of facts as interesting as I did. If you are interested in helping puffins, you can "adopt" a puffin through the Audubon Society's Project Puffin (http://projectpuffin.audubon.org/), a seabird restoration program. If you know of another nonprofit organization researching or helping puffins, please let me know about it!

—RA Anderson

PUFFINS

Types of Puffins

There are four species of puffins: Atlantic puffin, tufted puffin, horned puffin, and rhinoceros auklet. Three types of these puffins look similar, but the rhinoceros auklet has a much thinner beak and adults are more of an ashen gray color with a horn-like extension at the top of their bill. This bird has a fitting nickname: the unicorn puffin.

The Atlantic puffin (*Fratercula arctica*, which means "little brother of the north") is sometimes called the common puffin. They are a species of seabird in the auk (aka alcid) family that is most commonly found in Iceland. The Icelandic word for puffin is *lundi*, *lundy*, or *lundinn*.

Nicknames

Puffins have two common nicknames: "Sea Parrot" and "Clowns of the Sea." They get their nicknames for two reasons: first, because of their bright, clown-like facial markings and colorful beaks and feet; and second, because they are quite clumsy at their landings, crashing into one another or toppling head over heels onto the soft grass. In the ocean, they run across the surface to take off and belly-flop into rolls for landing.

Burrows, Not Nests

Puffins do not build nests on rocky ledges. They dig burrows high up on grassy cliff (*klettur*) tops. At about age five, the males come home and start digging their burrows . Within the next year or two, they will choose a mate and then both will finish the burrow and make it their home for the rest of their lives. Whether sprucing up an old rabbit burrow if the rabbits have left or starting a new one, they will both dig with their feet and beaks and gather grass, flowers, and other small plants to bring comfort to their burrow.

The Látrabjarg cliff located in the Westfjords is the westernmost point of Iceland and home to one of the largest bird colonies in the world. It is Europe's largest bird cliff at 8.7 miles (14 km) long on the coastline with cliffs 1,447 feet high (441 meters).

From Incubation to Adulthood

A puffin's egg takes about 36 to 45 days to hatch—this is called the incubation period. After hatching, the parents will feed their chicks by passing food from their beaks to their young pufflings. As pufflings age, their parents drop the fish so the chick can learn to feed itself.

Puffins take turns with their parental duties. One will stand guard and watch the burrow while the other flies off to catch fish for their puffling. When one returns, if needed, the mate will take off for more food. These parents work equally hard, teaming up for their little one to survive. Sometimes each parent must make eight to ten trips per day to find enough sandlance, or sand eels, to nourish their puffling chick.

Ordinarily, puffins lay only one egg per year, but research has shown that sometimes if an egg is "lost" early in the season, it is possible they can lay another.

About 30 to 60 days after hatching, pufflings are fully fledged, which means they have the feathers needed to fly and can live on their own.

When temperatures drop , they know it's time

to leave their burrows. Pufflings leave their burrows for sea at night, alone. Their fledging journey consists of bouncing and flapping their way down steep, rocky cliffs until they are safely swimming in the ocean. This takes place on a cold, dark night in late August or early September. Some pufflings become disoriented on their journey down the cliffs due to lights in the towns and end up landing in the middle of the towns instead of offshore. Once pufflings land on the ground, they are not able to take off again. In Heimaey, a southern Westmann island off Iceland, the children gather up the pufflings and take them home for the night. Early the next day, they release them to the sea. Once a puffling chick leaves their burrow, they must learn to fly, swim, and fish without their parents' help.

Pufflings don't return to land right away after the year they first fledge. They will stay out at sea for three to six years while they learn to hunt. When they come back, they usually return to the same colony where they were born to find a mate and start their own families.

As scientists continue to study puffins, they are having difficulties with tracking where they go when out at sea. Some successful geolocator data informed us that some puffins can travel as far south as the Mediterranean and over 5000 miles in the eight months out at sea. (Cited from *Puffins* by Drew Buckley)

When fully grown, a puffin is about 25 cm tall (about 10 inches) and weighs about 500 grams (1.1 lbs)—about the same as a can of soda. Male puffins are usually a bit larger than the females.

A puffin's average lifespan is about 30 years. The oldest recorded puffin lived to be 41 years old. Puffins mate for life and are devoted partners. It has been known that when a mate perishes, the other may die of heartbreak.

Preening

A puffin spends most of the day preening, which is when they spread oil from the preen gland located at the base of the tail all over their feathers, thus waterproofing their feathers. Their outer feathers have tiny barbs and hooked barbules that lock together. When their feathers are locked, they form

an almost airtight, life-jacket layer around their little bodies. Not only does this locked layer help them float, but it helps keep them warm while in the cold ocean for so long.

Underwater Skills

Puffins, like penguins (but not related), flap their wings underwater to gain speed or slow down, while they use their webbed feet as rudders to steer. When puffins swim, they look like they are flying underwater. However, before they dive, they must first release all of the air stored in their feathers. They force the air out by using special muscles under their skin.

A Day Fishing

Puffins will hunt 10 km to 110 km (6.2 miles to 68.35 miles) away from their burrows and make as many trips as needed to catch enough food for their families. Puffins can dive to a depth of 60 meters (200 feet) and hold their breath for up to a minute. The average dive is 20 seconds to catch an average of 10 fish per trip, but a puffin can fit up to 62 small fish at one time in its bill and will make as many as 10 trips.

Puffins in Flight

Puffins are not graceful fliers! They struggle to get into the air and then flap their wings 300 to 400 times a minute to just stay in flight. They can fly at around 48 to 55 mph (77 to 88 kph). Some say they look like black-and-white footballs in flight.

On Land and at Sea

Puffins are chatterboxes at their breeding colonies but silent while at sea. On land, these birds are super social and affectionate, but for the months at sea, they are alone and trying to blend into their surroundings.

When a puffin arrives back at the colony from being out at sea for seven or eight months, their sea legs aren't used to land. Some land just fine, but some bump into anything in their way, including a friend or their mate. I have seen puffins do head-over-heels somersaults and then, as they find their feet, shake it off like nothing happened and walk away in their unique gait, the Puffin Walk.

Puffin Walk!

Puffins' legs and webbed feet are set back on the body for swimming, so not designed the best for walking. On observation, they seem to have a few different gaits. They either duck their heads and waddle or stand tall and proud, marching in front of their burrows to protect their egg or puffling chick.

Their communication takes place in the way they walk. A low-profile walk is when a puffin walks quickly with its head down. They may be saying, "Just passing through, no worries." Or maybe, "I didn't mean to knock you down when I landed."

When guarding a burrow, a puffin may stand stiffly erect with its beak tucked in next to its body. They may walk with exaggerated foot movements or look like a soldier on guard duty, because they are. They are guarding their burrow.

Puffin Battles

Puffins rarely have issues with other puffins, but when they do... An aggressive encounter between two puffins may begin with tossing their heads and beaks up, which is called "gaping." The wider their beak is opened, the more upset the puffin is. They may even stomp their foot in place to show others their displeasure.

The rare occasion of a puffin battle brings quite the show. Several other puffins may gather around to watch a full-scale brawl. The two fighters will lock beaks and toss one another over as if in a wrestling match.

Puffin Colors

The Atlantic puffin is known for its outrageous orange beak and legs and webbed feet with small black claws. In the spring, its bill turns bright orange to attract mates, and in the fall, puffins shed the colorful orange outer parts of their bills as they return to the sea.

Puffins' winter colors are the perfect camouflage out at sea. Their feathers are black with ripples of dark gray and they have white underbellies, but the black and gray wrap around at points. Their white bellies make them look like the reflection of the sky to potential predators swimming below, and their black backs with the ripples of gray mask them in the darkness of the deep ocean, hiding them from predators above.

Puffin Beaks

Scientists believe puffin beaks are bright and colorful to attract a mate, but they have now discovered that they are fluorescent as well. So now we know that puffins can see fluorescence, but why they see it remains a mystery.

Puffin beaks are hinged in a unique way, enabling them to hold both top and bottom parallel. This allows them to hold several fish in a row, placing the fish spine on the edge of the beak. The placement of their catch is done when scooping fish—their grooved tongues gather the fish in rows by moving them back like a conveyor belt.

Puffins have nostrils on their beaks to breathe and also have a unique salt-fluid-draining gland near their eyes. Puffins drink seawater, and when a puffin shakes their head, these glands near their eyes release salt from their bodies.

Puffin Poop

Puffins do not poop in their burrows because it ruins the waterproof protection on their feathers, so they go to the front where they have made an area like a bathroom, or they go to the edge of the burrow to poop outside.

Penguins vs. Puffins

In *Puffins Encounter Fire and Ice* (Iceland: The Puffin Explorers Series Book 3), Árni and Birta overhear tourists calling them penguins at Diamond Beach. Even though puffins are sometimes confused with penguins, penguins do not live in Iceland. In the wild, penguins live almost exclusively in the Southern Hemisphere, although Galapagos penguins live just north and south of the equator, at the edge of both the Northern and Southern Hemispheres. Puffins only live north of the equator, in the Arctic regions.

Where to See Puffins

The best place in the world to see puffins is Iceland because many of the locations are accessible by a short walk from a car. Some other locations to see puffins from May–August are:

- Great Britain
- Mykines in the Faroe Islands, Denmark
- Røst, Norway
- Coastal Islands National Wildlife Refuge, Maine
- Witless Bay, Newfoundland, Canada

How Many Puffins in Iceland?

About half of all Atlantic puffins nest in Iceland, where the estimated puffin population is just that—estimated—because it is difficult if not impossible to count them with anything close to perfect accuracy. The National Audubon Society's Project Puffin estimates that there are 3 to 4 million pairs of puffins in the world and most of them can be found in Iceland, where about 60% of them—which would be approximately 1.8 million to 2.4 million—breed every year. Several Icelandic tourist organizations claim that 3 to 4 million puffin pairs breed in Iceland each year and the total number of puffins is 8 to 10 million birds.

So, the answer to how many puffins there are in Iceland is—too many to count! But we can be fairly sure that they number in the millions.

Puffins and People

Atlantic puffins aren't afraid of humans. They move cautiously around when humans are near but are more curious than scared. Puffins have been known to walk right up to humans and give them a closer look. They don't know that their biggest threat today *is* humans. Coastal development, tourism, oil spills, climate change, and the introduction of non-native predators like minks, cats, and dogs are all contributing to the decreasing number of puffins on the planet. But Árni, our *lundi* friend from Iceland: The Puffin Explorers Series knows that you will be a good friend and treat all puffins with care.

Protecting Puffins

People can unintentionally prevent *lundi* parents from performing their parental duties by simply standing around their burrows. A simple human touch could actually be very harmful for a puffin, ruining the special coating on their feathers and causing them to be unable to survive in the cold water.

Even though puffins are adorable and we would love to take them home with us, our ultimate responsibility is to protect them by leaving them alone, doing no damage to their habitat, and helping protect their environment by addressing climate change issues, both personally and as part of our local and world communities.

Puffin Predators

Puffins' main predators are great black-backed gulls and mink. Other predators are Arctic skua, Arctic foxes , rats, eagles, cats, and humans.

The American mink, a non-native, invasive species in Iceland, is a particularly vicious predator for puffins. The Arctic fox will hunt and kill one puffin for its meal, but the mink has been referred to as a "killing machine" because they will come into a puffin colony and kill hundreds of puffins at once, devastating the whole colony.

In Iceland, the Arctic fox is a main predator of the mink, so, in order to control the burgeoning mink population, humans have had to step in to protect the other species, especially seabirds, that the mink are devouring.

Black-backed gulls, the big bullies in the sky, are known to attack puffins arriving home with their catch, not only wanting to steal and eat their catch but to eat the puffin too! But puffins are clever. After

fishing, puffins arrive home as a group, which helps confuse the gulls. While the puffins are away, gulls have been seen at burrow entrances trying to catch pufflings to eat as well.

Puffins on Camera in Iceland: The Puffin Explorers Series Books

Although Árni and Birta have an adventure in this story that imagines what they might do with a day to themselves on their island, in real life, a puffling doesn't leave its burrow until the night he or she makes their journey to sea.

When they are hatched, they are a dull gray color, which helps them stay safe from predators. To help protect the pufflings, we didn't invade their privacy or disturb their burrows for photographs so we used photos of colorful adult puffins for our dear young friends Árni and Birta throughout the Iceland: The Puffin Explorers Series books. This is why little Árni and Birta do not look like gray puffballs in a dark burrow in those books. Thank you for understanding the importance of leaving wildlife as is, especially in this vulnerable time of their lives.

HISTORIC ICELAND

First Settlers

The first mention of Iceland may have come from the Greek geographer and navigator Pytheas, who described an island six days' sailing north of Britannia in 330 BCE. (Encyclopedia Britannica)

There is archaeological evidence indicating Gaelic monks from Ireland or Scotland (part of the Hiberno-Scottish mission) were in Iceland prior to any Norsemen/Vikings, possibly as early as the 700s, though not all scholars agree on that. They were known to travel spreading Christianity, but these might have been hermit monks. Then Vikings braved the uncharted seas in their Viking ships to find new lands. Whatever the reason, when "heathen" Norsemen and their gods began to settle in Iceland, the monks moved on.

According to *Landnámabók* ("Book of Settlements") written in the 1100s, Naddador, a Norse explorer from the Faroe Islands, accidently found Iceland when he went off course and drifted into the island in the mid-800s. When he came ashore, it was snowing, so he named this land Snæland, which means "Snow Land."

A few years later, a Swedish Viking named Garðar Svavarsson spent a winter in Iceland and named it after himself: Garðarshólmur (Gardar's Island).

Then, a Norwegian Viking named Hrafna-Flóki Vilgerðarson was the first Norseman to sail to Iceland intentionally. Hrafna-Flóki changed the name from Gardar's Island to Snæland, and later changed it to Iceland after spending a couple of winters there and observing the fjords full of spring ice. A popular folk tale says he named it Iceland because he wanted to mislead his enemies so they wouldn't follow him and take over his land, and that he gave Greenland, which is normally covered in thick sheets of ice, its name to scare off anyone from going to Iceland because if a place named Greenland is shrouded in ice, how cold and icy must a place called Iceland be?

It's a memorable story, but the story about the ice-laden fjords may be closer to the truth. Or maybe it's the story documented in *The Saga of Erik the Red*, which says that Erik, who had been exiled from Iceland for committing murder, sailed westward to Greenland, named it, and was the first colonizer of Greenland around 985. According to the *Saga*, Erik's reasoning went like this: "He called the land which he had found Greenland, because, quoth he, 'People will be attracted thither, if the land has a good name.' And so it is Greenland to this day." But the truth is that we can't really be sure. People have always told stories—folk tales and myths—to explain things we can't actually know or fully understand, so we continue to study them to find out more.

It wasn't long before Ingólfur Arnarson arrived in Iceland after he and his stepbrother, Hjörleifur, had killed two sons of a Norse earl. The Norse earl took all of their land. In search of new land, they sailed to the west to start over. *Landnámabók* ("Book of Settlements") explicitly states that in the year 874, Ingólfur Arnarson became the first permanent Norse settler on Iceland.

Ingólfur Arnarson didn't claim the first land he arrived at; he wanted his gods to choose for him. The pillars on his boat were carved to represent his gods, so he tossed them into the ocean, claiming that his gods would choose the land where he would settle. They spent three years in a temporary location while waiting for the two slaves he'd sent out looking for these carved pillars. The slaves reported back and led them to the pillar gods' location. Ingólfur Arnarson named

this area Reykjavík, meaning "steam bay" or "smoke bay," and rewarded the two slaves with their freedom and land. Reykjavík is now the capital city—the only city—in Iceland.

Populating Iceland

From 870–930, between 30,000 and 40,000 people settled in Iceland. Forty percent (40%) of Iceland was covered with vegetation when they first arrived. Then birch trees, willow trees, and smaller native trees were all cut down for construction and other uses. Now, only 25% of Iceland is covered by vegetation and the only trees are in private, small groves planted by landowners, typically in fenced areas. It is hard to grow trees where there is fire and ice, but it is not impossible. For the past couple decades, greenhouse farmers have been growing tree seedlings in green houses and moving them outdoors when they grow large enough.

There is a joke the locals tell: If you get lost in a forest in Iceland, just stand up.

Vikings and Norsemen

Norsemen and Vikings were a Germanic people (meaning they spoke a Germanic language—in this case, Old Norse) who lived in Scandinavia. Viking people were variously called Northmen, Normans, Danes, Sus, or Varagians.

Vikings lived in the same place and spoke the same language as Norsemen but did different things for a living. Norsemen made their money by trading, so were sometimes called Norse traders. Vikings were farmers , unless something came up that required warriors, like raiding a new land to claim it for their own or pillaging a village or monastery to acquire goods for themselves, and then they were fierce fighters. Not all farmers participated in raiding—mostly the younger ones, while the older ones stayed home to tend the farms.

The Viking Age effectively ended when

the raiding stopped—the year 1066 is usually cited. One theory is that the arrival of Christianity in Northern Europe was a factor in persuading the Vikings to give up raiding. Their descendants, mostly in Iceland after the end of the Viking Age, created masterpieces of poetry and sagas in their own language, which form much of the basis of our knowledge about them and their times.

Vikings' first *spoken* language was Old Norse, but their first *written* communications used symbols to represent things like ships and horses and convey their beliefs, power, protection, and magic.

Viking in Old Norse was *vikingr*, which means freebooter, sea-rover, pirate, and Viking. It also explained that they were people who came from the fjords, or *vik* areas. A fjord is a narrow and elongated sea inlet with steep land, or cliffs, on three sides, also known as gateways. The word *vik* means a creek or an inlet or small bay.

Researching the Language for Iceland: The Puffin Explorers Series

This was one of the most difficult things I had to do! There are few places to find English-to-Icelandic translations besides studying at a university. I purchased books and visited websites and then found that all the information, including the books, seemed to conflict! It remains a mystery to me where to find American English pronunciations for these words without memorizing their alphabet, their unique letter sounds, and just plain trying my best.

I didn't give up looking for help with the Icelandic pronunciations though. Thankfully, most Icelandic people also speak English as a second language, so I reached out to a friend in Iceland who helped me make sure I wasn't using the wrong words.

The Icelandic language, *íslenska*, is well preserved, and Icelanders, *íslendingar*, are very proud of this fact. A Bible written in *íslenska* in the 1500s can still be read by Icelanders today.

To me, most Icelandic words look like my cat walked over my keyboard, laid down, and continued to mash the keys until she heard someone opening a can of tuna and ran off.

An example of my theory is the word

Vaðlaheiðarvegavinnuverkfærageymsluskúraútidyralyklakippuhringur,

which is the longest word in the Icelandic language that I know of. Please don't ask me to pronounce it. This word supposedly means, "key ring of the key chain of the outer door to the storage tool shed of the road workers on the Vaðlaheiði."

Amazing, isn't it?

Turf Buildings

Turf houses may have been the original green buildings of the world. In Iceland, turf buildings date back more than 1,000 years. The first settlers of Iceland, around 870 CE, had to build with materials such as turf (sod) from the wetlands and lava stone because of an extremely limited supply of wood and lack of lime in the ground for making concrete. They used special hand-made tools and a lot of hard labor. Upper-class people and poor people all used the same building materials that were locally available, from the earliest Viking settlers to the mid-1960s when building supplies were being shipped in and more modern homes were being built.

Each structure was built by digging it into the ground, which protected it from the elements even more. The wall layers were turf from the wetlands, then they added a layer of lava stone shaped in squares with flat edges, and then turf again. The turf was "hammered," or shaped down, by packing it tight with unique handmade tools. Then they would trim the turf edges. This process would be repeated until the walls were as tall as needed. Roofs were steep, and

another layer of turf was used to prevent leakage inside. Horses were used to build turf houses and barns as late as 1965 because there were no tractors.

These structures were built in clusters to share warmth and walls and were connected with a small passageway. The main living quarters housed all of the people, with beds lining the outside walls and tables in the middle. It was where they gave birth, conceived children, and died—everything in one room. Other attached buildings were for kitchens, barns, food storage, and livestock.

These structures generally lasted for 30 to 40 years, then each home and structure was recycled and reused from generation to generation. Older structures still standing are often used for barns and museums. The last turf houses were built in the 1970s .

Lighthouses

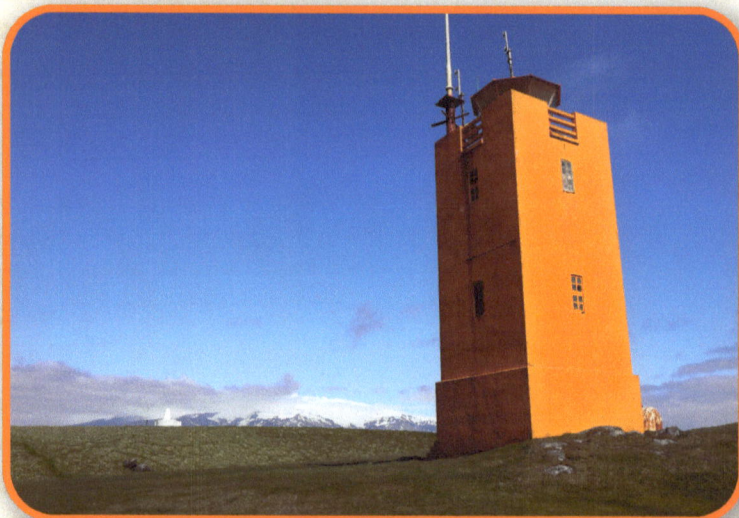

Lighthouse is *vegagerðin* in Icelandic. They are guiding lights serving as navigational aids to warn or lead ships around hazardous or dangerous areas or into ports and basins. They act as traffic warning signs for seamen. There have been approximately 120 lighthouses built in Iceland, and most are still standing and active. As of this writing, 104 are active and managed by the government.

There are a number of seal and harbor beacons scattered around Iceland as well. These are the harbor lights that float anchored to the ocean floor and help guide vessels through the inlets. They are called seal beacons because seals hang out on the bottom platforms.

In 1878, the Valahnúk was the first lighthouse built, located on the Reykjanes peninsula, but was destroyed in 1905 by an earthquake. It was later replaced by a new lighthouse, the Raykjanesviti . In 1908, when the new lighthouse was finished, the Valahnúk was demolished.

The oldest working lighthouse is Gróttuiti, built in 1897 on Grótta Island, which is connected to

Reykjavík, the city of Iceland.

The Dyrhólaeyjarviti lighthouse was built in 1927. Located just north of the town of Vik, it overlooks Dyrhólaey's black beach and the unique views of the peninsula.

In the Puffin Explorers series, Árni's family lives near the twenty-foot-tall Bjargtangar lighthouse built in 1913, which sits on the Látrabjarg cliffs on the westernmost point of Iceland, overlooking the Westfjords peninsula and uses, or is marked by, three white flashes every 15 seconds.

Faeries, Trolls , and Elves

Some Icelandic people still believe in Huldufólk—the "hidden people"—which are elves, trolls, and faeries who live within the beautiful landscapes and wreak havoc on those who disturb their homes and communities. It is understood that you should respect the Huldufólk or they will curse you. This is why you shouldn't throw stones in Iceland, because you might hit one of these legendary magical creatures. Several stories from road engineers encountering huge obstacles during road construction have arisen, and to carry on peacefully, they have had to move boulders up to 50 tons—sacred spaces of the faeries or elves— to a safe location so the Huldufólk would be happy and allow construction to continue without harm or foul.

Troll's Lava Pool

Icelandic folk tales have been passed down from generation to generation. While in Iceland, I visited Brimketill Lava Rock Pool, "Oddny's pool," located on the peninsula of Reykjanes. This lava pool was named after a famous troll named Oddny who had several children to feed. Because troll's would turn to rock if the sun kissed their skin, she had to hunt or fish in the dark of night. She came home early morning after her hunt all stinky and exhausted, so she decided to wash and soak in her private pool on the peninsula. However, Oddny fell fast asleep while soaking in her pool. Dawn woke, but Oddny snoozed on, meeting her fate as the sun turned her to stone. Since that day, Oddny has washed away, but the pool remains to this very day. When you feel the mist of the ocean sprays, it's Earth's way of remembering Oddny. About 30% of Iceland is covered in lava fields. Unique lava formations like Brimketill Lave Rock Pool are scattered all across Iceland, but none as special as this one.

Trolls and Puffins as Friends

The clever book *Icelandic Trolls*, written and illustrated by Brian Pilkington, gives us a whole new, lighter perspective about trolls. One section tells of some trolls who hollowed out a space behind the puffin burrows, within the mountains. Friends of the puffin parents, the trolls help take care of the eggs and then the hatched pufflings from May to

September—the entire time puffins spend on land. On this positive note, we get a new vision of trolls as puffling guardians.

If you ever have a chance to visit the Blue Lagoon in Iceland, take the walking tour to hear the story of the Huldufólk who halted the expansion until it was approved by a faery. They have proof in the walls.

The locals say, "The sulfur you smell in certain areas of Iceland is the Huldufólk's dirty bathwater."

Recommended troll books:

- *The Legend of the Icelandic Yule Lads* by Heidi Herman
- *Guardians of Iceland* by Heidi Herman
- *Icelandic Trolls*, written and illustrated by Brian Pilkington

FARMS

Farmers' fodder crops (grass crops) grow extremely well in Iceland. The long hours of daylight, cool days, and lack of insects and pests are all beneficial to growing an abundance of hay all summer long. Iceland has less air pollution and less ground and soil pollution, which also helps the hay grow and makes it very healthy.

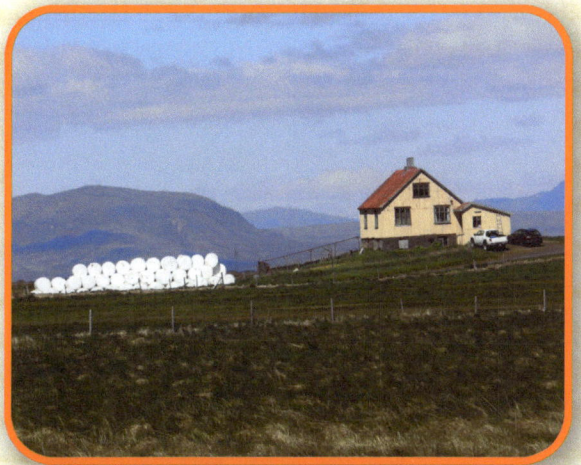

Farmers cut, dry, roll, and then wrap the grass hay tightly in plastic . It must be wrapped perfectly with not one hole, or moisture will make the whole 1,200-pound bale moldy and useless. Farmers use colored plastic, originally to blend in with the landscape, but more recently, they've been using colors such as pink for cancer awareness.

When driving past farmers' fodder crops, the rolled bales look like candy mints and marshmallows. Icelanders like to tell the tourists that farmers leave these mints and marshmallows out for the trolls to eat to make sure they are well fed and happy.

The cost of the grass/hay-roll bales in Iceland varies from year to year. In 2019, the cost of a 1,200-pound bale was 12,000 ISK (Icelandic Króna), which at the time was approximately $120 US dollars. It costs farmers' approximately 80,000 ISK ($80 USD) per bale to grow it, cut it, roll it, and wrap it in plastic.

In the United States, the same size round roll bale would cost a farmer approximately $60 to $90 USD to produce without the plastic protection wrap. Typically, the US rolled-hay is not the same quality as the hay produced in Iceland. The sales price for a roll bale of the equivalent size (but not of the same quality) of the ones in Iceland would cost between $90 to $120 USD. The quality of Icelandic hay is much healthier than the United States because in Iceland they do not use pesticide treatments, making their soil healthier.

In a high-demand season like winter or a year that presents difficult conditions, prices can be much higher. For instance, Norway had a bad hay-growing season in 2018 due to weather and needed to import hay from Iceland. Iceland's prices per bale could have tripled if the farmers wanted, as is the typical effect of supply and demand. However, while staying at one of the farms in Iceland for a couple days, I asked the owner how much hay was being sold for to the Norway farmers. The answer amazed me in respect to how caring everyone seemed to be. This farmer was so concerned over the well-being of the animals in Norway that they sold their hay at cost to farmers who agreed not to sell it and only use it for their livestock. The farmer in Norway paid for the shipping of the bales on top of the cost per bale. I couldn't say if other farmers did the same, but this one had a heart of gold!

Icelandic Horses

Icelandic horses are descendants of Viking horses brought to Iceland around the 800s (ninth century). They were shipped over from Europe's mainland by boat, so they needed to be small and sturdy and have nice dispositions.

After the settlement period, Icelandic horses were used for farm work. People who owned a horse were able to travel from the country to the towns' trading posts, allowing the farmers to make more money. Owning a horse represented family wealth.

Icelandic horses have the ability to cope and adjust to extreme weather conditions without being put up in barns for shelter. This makes them a perfect horse breed for the challenging weather in Iceland.

Their average size is about 14.2 hands (4'9" tall), measured from the bottom of their hoof to the top of their withers. In the US, we consider 14.2 hands to be the size of a large pony. However, to call an Icelandic horse a pony is taken as an insult by Icelanders because they are justifiably famous for their Icelandic *horses*.

Icelandic horses have an additional, unique fifth gait. This four-beat, rhythmic gait is smoother than a trot. When going slow, it is referred to as the *tölt*, but it can accelerate quickly to a much faster speed, called the *skeið*. Known for being a comfortable gait with rapid acceleration, it was believed to have developed due to Iceland's rough terrain and would have been handy in battles between clans. However, *ScienceNordic*, the English-language source for science news from Nordic countries, explains the science and history behind the special gaits: "The mutation [that made the new gaits possible] occurred in the DMRT3 gene, affectionately known as the 'gait keeper' gene, which controls the horses' leg movement. This mutation led horses to develop new gaits, such as the Icelandic *tölt*, which is so smooth that the rider appears to be almost completely still while the horse is in motion." *ScienceNordic* also reports that new research "provides clear evidence" that the Vikings were expert horse breeders who bred particular traits in their horses to make these gaits possible.

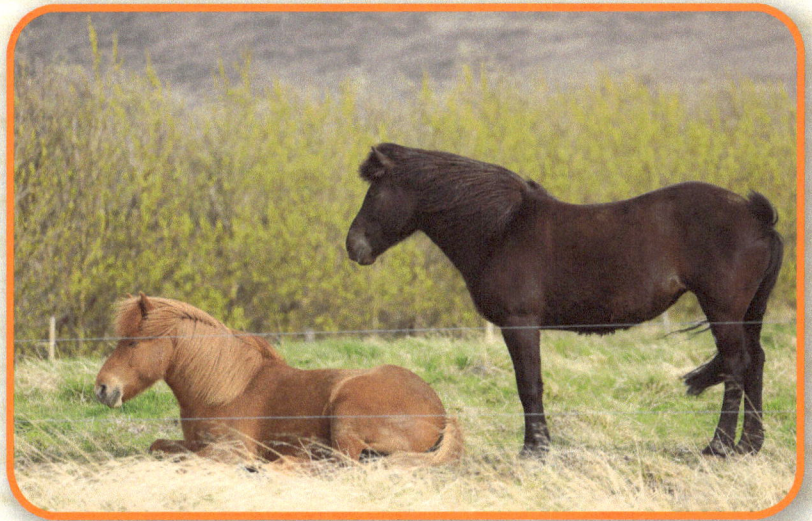

More Icelandic horses live outside Iceland than inside Iceland. Once an Icelandic horse leaves the country, it can never return because the native breed is believed to be susceptible to foreign diseases and infection could devastate the Icelandic horse's population. Currently there are approximately 80,000 Icelandic horses in Iceland.

Lambs and Sheep

There are more sheep in Iceland than people! There are 800,000 sheep and only 337,780 people as of September 2018.

A female sheep is called an ewe, and male sheep are called rams. A dam is a mother sheep. Icelandic sheep are of medium size, with mature ewes weighing an average of 150 to 160 pounds and rams weighing an average of 200 to 220 pounds.

The ewes usually birth in May and have two lambs, which is why you will mostly see them in groups of three. Sheep are fast on their feet and have little flocking instinct, so in the summer, they can be

found anywhere except on the glaciers. Ewes have excellent mothering instincts, bond quickly with their lambs, and have plenty of milk to feed them. Lambs grow extremely fast, sometimes gaining twenty pounds a week. Even though sheep are good mothers, they don't often stay near others in the herd. The lambs seem to wander off from their mother often. When they notice, they start baaa-ing like crazy, and as soon as they hear their mother call out, they race off to find her and go straight to her udders as if they had missed a day's worth of milk.

Genetically, Icelandic sheep remain unchanged after being in Iceland for over a millennium. They arrived with Iceland's first settlers from Norway about 1,100 years ago, their wool and meat making it possible for people to survive in the harsh, cold conditions. Icelandic sheep have been the lifeblood of Iceland for centuries as a meat, dairy, and fiber breed. Their meat is high in Omega-3 and fatty acids. They give great milk for a variety of products such as *ís* (ice cream), cheese, yogurt, and soap. Their fiber (wool) is warm, dense, and water-resistant. They are one of Earth's oldest, purest breeds of domestic sheep.

The reason Icelandic lamb tastes as delicious as it does is rather morbid. As they are free-roaming throughout summer, the sheep graze on Icelandic thyme, unwittingly flavoring their meat as they graze the hillsides.

Iceland's Sheep Rule the Road!

Icelandic sheep are fast and sturdy but have little flocking instinct, so they roam and spread out across the land, including the steep cliffs. Because they are a triple-purpose resource, they are well protected by laws in Iceland. They are out free, grazing the land from after they birth their new baby lambs in May until September. It's true, and it's the law: cars must yield to sheep in the road! Sheep in the middle of the roads are common, and yes, if you hit them, you must pay for them.

Tourists are discouraged from stopping in the middle of the road and getting out to take pictures for several reasons. You might not be the only car using the road and could cause traffic issues, but it can also cause the farm animals extra stress and disrupt their normal daily routine.

Rettir: Gathering Sheep for Winter

All sheep will stay on their own farms from September to May, or at least until after they have birthed their lambs. In September, the sheep are rounded up by people riding horses, with the help of sheepdogs, then by foot where the terrain does not allow riding, and they call this process *Réttir* (corral). Sheep can be found anywhere in Iceland except on the glaciers, so to gather them all may take farmers, sheepdogs, and people with horses . It

takes days—sometimes weeks—to get them all home. The sheep are identified by ear tags or markings, thus making it easier to know which sheep belongs with which farm.

Réttir is a very popular festival celebrated throughout the country in September. Many nonfarming Icelanders go to help and take part in the festivities (finding sheep) and the festival that follows. Recently, some tour operators have started offering tours to *Réttir* in September, which gets the tourists involved with the roundup as well.

Multi-Purpose Farms

There are farms with several purposes. Some have a restaurant, rooms for rent (called a farm stay, they are like a hostel or motel), horseback riding, baby goats and pigs to adore, and cattle for their dairy products and homemade ice cream .

Icelandic Sheepdogs

Hundreds of years ago, farmers brought over dogs to help herd and guard their sheep. The Icelandic Sheepdogs were developed from their Nordic cousins but are smaller. They are intelligent, small, fast, protective, and loyal working dogs.

Other dogs and animals being brought over to Iceland brought diseases, so Iceland banned other dogs from entering the country in the late nineteenth century when the Icelandic Sheepdog faced threats of extinction. Later, pet vaccines were introduced, and since then, the population of the Icelandic Sheepdog has recovered.

Fleece, or Wool

Fleece is their coat, their fur. It is soft and insulating, providing great protection against the cold in Iceland. One of Iceland's largest industries is exporting wool and wool products.

Lopi is a mix of the *tog* and *thel* fibers. It is durable, warm, lightweight, breathable, and water-repellent, and is best known for its use in making socks.

Tog is the coarser, longer, outer fleece that is mainly used for weaving. This is the layer that protects sheep from harsher weather and keeps them dry. After being sheared from the sheep, the tog's main use is for weaving.

Thel is the softer, thicker fleece near the body of the sheep. These soft fibers are used for undergarments and baby clothes.

Fleece's natural colors range from white, ivory, brown, taupe, silver, blue-gray, charcoal gray, to black. No dyes are needed, but the white fleece will hold dye well when colored. "Fleece as white as snow" only happens if snow is actually on top of them.

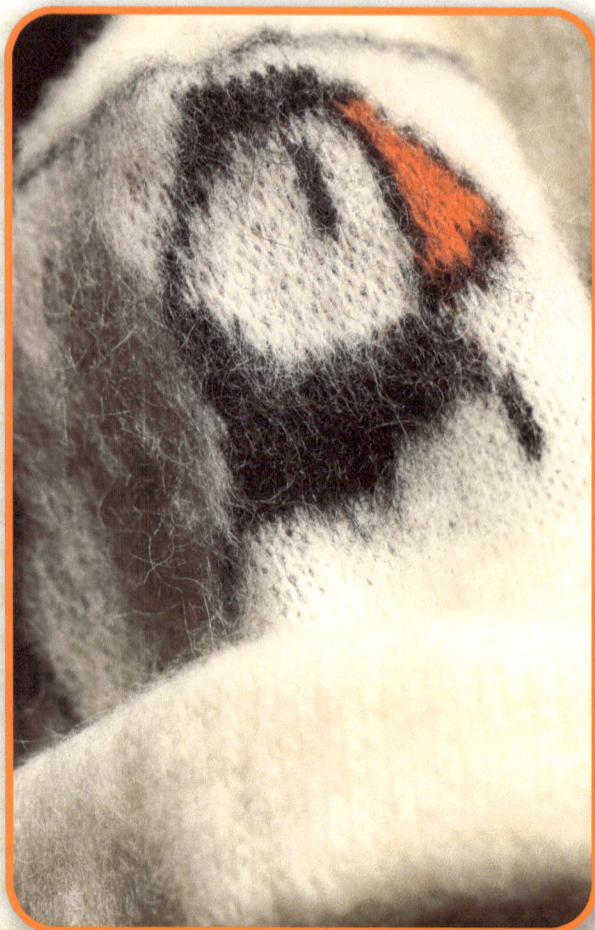

WILDLIFE

Whooper (pronounced *hooper*) swans fly nonstop from Scotland to Iceland during spring migration. A pilot once reported seeing a flock of whooper swans flying in a V formation at their plane's 8,000-foot cruising altitude. Scientists have recorded these swans flying at altitudes up to 27,000 feet and at an average speed of 60 miles per hour.

Whooper swans from Scandinavia, northern Russian, and northern Asia spend their summers breeding in Iceland. Some areas in Iceland are rich in iron compounds, which stain some of their head and neck feathers a rust color, but this is temporary because the color is lost when they molt.

Whooper swan families are known to stick together and make the big migration together. They have a single mate their whole lives, and both parents care for their cygnets (their babies), which stay with their parents through the winter and migration in the spring. They also return to the same nesting areas over the years.

The whooper swan is one of the largest swans and has yellow and black on its bill. The scientific name for a whooper is *Cygnus cygnus*, and they are classified as waterfowl. Their average lifespan is only nine years. They weigh an average of 9.3 kg (20 lbs) and have a wingspan of 2.3 m (7.5 feet).

Other Birds in the Wild

Other birds outside of the aquatic areas are skuas, gyrfalcons, golden plovers, snipes, barnacle geese, Eurasian Oyster catcher , and ptarmigans, all of which call Iceland home. Besides being one of the world's most wide-ranging birds, the raven makes a showing in

Iceland as well. They also play an important role in Icelandic folklore and pagan beliefs.

Arctic terns, great skua, and sea eagles are found around the coastal waters along with the gulls. Oh, the gulls! As you know, great black-backed gulls hunt puffins, and so will the great skua. Arctic terns eat the same fish as the puffins, but these little 3 to 4.5 ounce birds will not bother puffins, though they are also hunted by the great black-backed gull. Both the herring gull and Iceland gull eat other birds and their eggs, and the glaucous gull preys on small birds. There are other gulls as well, like the lesser black-backed gull, the common guillemot (common murre), the ivory gull, the herring gull, and a few more not listed here. Most gulls are bullies and will do just about anything from stealing food to killing for their food and then eating their kill.

Lake Mývatn : A Warm Winter Home for Birds

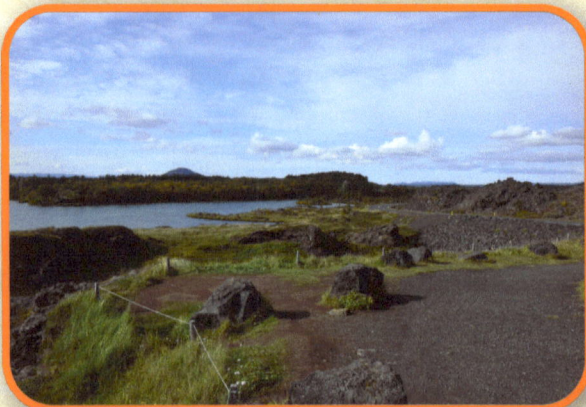

In Iceland, the volcanic lake Mývatn is known for birdwatching, heated lagoon water, and views of the northern lights. The lake never freezes, and it attracts about 85 species of birds, including harlequin ducks, common scoters, whimbrels, Arctic terns, Barrow's goldeneyes, goosanders, mallards, barnacle geese , and more. Lake Mývatn is home to fourteen different species of duck.

Even in the winter you can find birds, including some whooper swans. However, whooper swans are a migratory bird: typically, most whooper swans will migrate to their southern homes. Whatever the reason these birds may not be able to migrate south, they are safe over the winter in the Mývatn lake and River Laxá areas, where the water is warmed geothermally by nature so it doesn't freeze and there is an ample amount of vegetation year-round for them to eat.

Arctic Foxes

The Arctic fox was the first land mammal in Iceland, stranded 10,000 years ago at the end of the Ice Age when ice broke free, disconnecting from other land formations.

The coat of white Arctic foxes changes from snow white to brown in the summer to help them blend with their habitat. The blue fox doesn't change colors, but the sun bleaches their fur over the summer, helping them camouflage themselves in the winter. The Arctic fox has the warmest fur of all animals. Their coats can keep them warm even in subfreezing temperatures of minus 76 degrees Fahrenheit. They live off eggs, birds, berries, and now they prey upon other critters humans have brought over to this nation, like mink and small farm animals.

Minks

In 1931, the American mink (*Neovison vison*) was imported to Iceland for fur farming. Quite a few animals escaped, and the first den in the wild was discovered in 1937 in the capital area. Within 35 years, they had been spotted all over Iceland.

Minks are carnivorous, meaning they are meat lovers like lions and tigers. They are semi-aquatic, part of the family *Mustelidae*, basically meaning that they are cousins to weasels, otters, skunks, and ferrets. The mink in Iceland can travel at speeds of around 4 mph, weigh only 1.1 to 1.8 pounds, and are 13 to 18 inches tall. They live in denning sites close to water in well-maintained bank burrows. They are good climbers and swimmers, able to dive down approximately 16 feet in the water. Mink only live a very short time—6 months to 3 years.

At the peak of demand for mink fur in the 1980s, there were 240 mink farms in Iceland. There are still mink farms in Iceland today, but the ones that remain are operating at a loss, meaning they aren't making any money. Only time will tell what happens to the mink farms.

Reindeer

In the 1770s, reindeer were brought over to Iceland from Norway. Farmers didn't want to farm reindeer, so they released them into the wild. There are approximately 6,000 to 7,000 reindeer living in southeastern and eastern Iceland. They blend into the landscape so well that people often don't see them grazing off the side of the main road.

Sea and Sea Life

Sea currents in Iceland are extreme, and the undercurrents are deadly. An island in the North Atlantic Ocean, Iceland borders the south of the Arctic Sea. The Gulf Stream current swirls around the southern and western parts of Iceland, meeting up in the north and merging with the cold, east-flowing currents of the East Icelandic current and East Greenland current. This may be the cause of some interesting and extreme sea currents and activities.

Warning signs to use extreme caution are located in areas where tourists may walk near or to the beaches. What people don't know is that every eighth wave or so can be huge and easily carry an unsuspecting person away. Undercurrents are dangerous because the water below the surface is moving in a different direction than the surface current, away from the shore, sweeping out to sea. Fear, panic, exhaustion, and Iceland's cold sea temperatures are some of the reasons these undercurrents have claimed several lives. When in Iceland, pay attention to the warning signs!

Seals

The two species of seal found on Iceland's shores permanently are the harbor seal and the gray seal. These seals are spotted nearly everywhere around Iceland, but the best places to see them are the Westfjords, the Vatnsnes Peninsula, the Snæfellsnes Peninsula, and the Jökulsárlón glacier lagoon.

Four other species of seals visit Iceland on occasion: the harp seal, bearded seal, hooded seal, and ringed seal. Another friend making a very rare visit from time to time is the walrus. Walruses once had a large population in Iceland, but in the early seventeenth century, they were hunted nearly to extinction.

Long before humans came to Iceland, seals inhabited the cold, fertile waters and the long stretches of rocky coastal shores year-round and large seal colonies evolved. When settlers arrived, seals were an essential resource for their survival. They were able to hunt seals in great numbers due to the lack of fear seals had for humans. Seals provided the people with

food, clothing, and oils, but the people became greedy, killing for fashion, and seal numbers dwindled. Today's seal-watching industry has boomed, and their population has become more stable, especially since the opening of the Icelandic Seal Centre in Hvammstangi, which is dedicated to researching and raising awareness about seals.

Seal pups are born around the month of June, already with fur and ready to swim.

Dolphins and Whales

There are over twenty different species of whales and dolphins in Iceland's fertile sub-Arctic waters. Whale species seen regularly on whale-watching tours include mink, fin, blue, humpback, sperm, and killer whales. *Whale* in Icelandic is *Hvalur* (kva-lur). Most whale-watching outings include seeing dolphins and harbor porpoises showing off in the bow waves of the boats.

The whale-watching industry is changing the way Icelanders view whales because they are earning more money showing people whales than fishing for whales. Whaling started in the early 1900s, and most whaling was stopped due to the International Whaling Commission Treaty. However, this treaty, signed by almost every nation, isn't a law but rather more of a formal agreement not to kill whales. The only nations allowing the killing of whales are Japan, Norway, and Iceland. Unfortunately, endangered blue whales and fin whales are amongst the whales killed. Whether to allow whaling or make it illegal is an ongoing debate among country leaders.

Mink whales are eaten by visitors. Yes, many tourists are under the impression that eating whale is a tradition in Iceland, but it is not. Most whale meat on the market is consumed by tourists, not locals.

Polar Bears

Polar bears also do not live in Iceland. They live in the ring around the Arctic Circle, in such countries as Canada, Alaska, Russia, Norway, and Greenland, all in the Northern Hemisphere where warming ocean temperatures are negatively affecting their habitat and food supply, as is also true for penguins and puffins.

Years ago when there was more floating ice, on the rare occasion, a polar bear would float over from Greenland. The polar bear would arrive starving and attack anything upon arrival. Polar bears are one of the few animals known to hunt humans, so, considering this and that the amount of money it would cost to capture and return the polar bear to its home is estimated at $75,000, unfortunately, any polar bears wandering into Iceland are killed upon arrival. The last polar bear seen in Iceland was in the summer of 2016.

EARTH, WATER, FIRE, AND ICE

Layers of the Earth are complex, but the four basic layers are the crust, mantle, outer core, and inner core.

The crust is our home—the continents and the oceans. This outermost shell of Earth is broken up into sections called tectonic plates that are basically made of granite and volcanic lava called basalt. There are seven major plates and eight minor plates. Two of the seven major plates—the Eurasian and North American plates—meet and run through Iceland.

Most of Earth's mass, the mantle, is made of a really thick layer of hot, molten rock called magma that moves slower than thick asphalt. This is also where most of Earth's heat is stored. This moving, hot rock can cause the tectonic plates to move. When this movement occurs, it can cause several different things to happen. Two examples that are most common in Iceland are earthquakes and volcanic eruptions. They can happen together or one right after the other.

The outer core consists of iron and nickel so incredibly hot that they are in liquid form.

The inner core is the hottest. Made of iron and nickel, it is condensed into a solid state by the Earth's gravitational pressure.

If you took a hard-boiled egg and were able to crack the shell so it has seven large sections and eight smaller sections, it would give you an idea of how the Earth's plates look. Then, if you put pressure on one section of shell, that section would bump into or rub up against another section. Of course, no hot magma would rise nor would eggquakes happen, but this was always a good way for me to picture how tectonic plates move around our Earth.

Volcanoes

It is true that Iceland's volcanoes are active! In fact, Iceland has 125, with approximately one eruption every four years. It's the most active volcano area in the world, and four of their volcanoes, as of this writing, seemed ready to erupt. These Iceland volcanoes are Bárðarbunga, Katla, Hekla, and Grímsvötn. The largest and possibly most dangerous volcanic eruption would be Katla, but the volcanoes "hiding" under the great glaciers could cause a great melt and flooding across the island.

Two of Earth's tectonic plates meet and run through the island of Iceland, and where the plates meet, they can shift, crash, and pull apart, causing earthquakes. Earthquakes and volcanic eruptions can follow each other, making it difficult to detect what will happen next, but an earthquake is an additional indication of how close magma is to the surface and how close it is to an eruption.

Only a few miles from the Dyrhólaey lighthouse in south Iceland sit three of Iceland's volcanoes that are past their predicted eruption dates. What happens to the area, the animals, and the people when a volcano erupts depends on the size, or magnitude, of the eruption.

Scientist are actively surveying and studying the earthquakes and volcanoes in Iceland 24/7 because so many volcanoes are ready to erupt and earthquakes and volcanic activities can be related. The earthquakes coming from Bárðarbunga volcano are a result of trapped magma movement within the magma chambers. When surfacing, it travels through a volcano vent and then is called lava. The magma from Bárðarbunga volcano is causing

excessive pressure on the rocks above the chambers, causing the earth to quake. This is why scientists believe Bárðarbunga volcano may be the next to erupt.

Surtsey

Under the ocean near the Earth's crust, tectonic plates can pull apart due to the movement of slow-moving magma building underneath. When the plates pull apart, releasing magma can be like an underground volcano erupting. These can be explosive eruptions able to create an island, or they can be slower, effusive eruptions that slowly build up, and the results won't be seen for thousands of years.

Surtsey, a volcanic island, is one of the youngest islands in the world. Named after Surtr, the fire god from Icelandic mythology, it was created by a volcanic eruption 130 meters (427 feet) below sea level. Ash and gasses blew about four miles high, raining ash on the nearby Westman Islands. The eruption started at the end of 1963 and lasted until 1967. The island has been legally protected since birth to keep it free from human interference. It looks like a private bird and seal sanctuary, but it is also a long-term biological research site for programs conducted by Icelandic and American scientists. It has provided an invaluable opportunity to study primary succession, supplying a unique scientific record of how plants, animals, and marine organisms left to their own devices colonize a newborn, virgin piece of Earth.

When new land is formed, such as an island made from lava, it is a lifeless area with no soil and is incapable of sustaining life. Primary succession starts with this lifeless area and can take up to 1,800 years for an ecosystem to form. Wind, water, and pioneer species help weather the rock to form soil. The first stages begin with pioneer microorganisms. About 20,000 species of microorganisms are referred to as lichens. Lichens are the first to colonize on bare rocks, where they produce and ooze acids over the rocks, which helps break down rock to start the soil-production process. Algae provides energy for the microorganisms and fungi captures the moisture that *most* organisms need to survive.

Surtsey has been designated a World Heritage Site by the United Nations Educational, Scientific and Cultural Organization (UNESCO). This organization can declare a historical building, city, island, lake, monument, wilderness area, or other important site or place a World Heritage Site if they think the landmark is at risk of being ruined by humans. A

UNESCO landmark is protected by international treaties.

The first puffins nested in Surtsey in 2004.

Laki

The Laki volcano erupted on June 8, 1783, and the effusive eruption continued for eight months. Iceland's valleys were filled with lava, and it was estimated at 330 feet deep, covering 965 square miles of surface area.

The ash poisoned the land and the sea, causing all of the area's farm animals to die. People had to leave in a hurry so they wouldn't die as well. Way back in 1783, many people couldn't travel quickly enough to get away, so about 9,000 people were killed by the Laki volcano eruption.

Currently, ancient moss covers the lava fields from the Laki eruption. It's a beautiful and fascinating sight.

Hekla

The stratovolcano Hekla is one of Iceland's most active, and it is due to erupt at any time. Hekla is a stratovolcano because it has built up, shaping into mountain form, over thousands of years with layers and layers of hardened lava, tephra, and ash. Hekla has had explosive and effusive eruptions.

Eyjafjallajökull

Eyjafjallajökull erupted in April 2010, and it was explosive! The meeting of one body of magma (common volcanic rock basalt) and a slightly different magma (made of silica-rich trachyandesite) caused super combustion. Ash and other particles exploded into the atmosphere, grounding airplanes. Over 95,000 flights were canceled around the world due to ash floating in our atmosphere. Locally, no one died, but roads, homes, and farm crops were damaged and there were floods from the glacier melt. But because of the air traffic situation, this volcanic eruption was one of the most famous in Iceland.

Holuhraun

The Holuhraun volcanic eruption started at the end of August 2014 and lasted until the end of February 2015. Continuous fountains of boiling molten rock produced miles and miles of lava fields, the largest in Iceland since the Laki lava field was created in 1783. However, this new lava field is not located in a populated area, so people for the most part were safe and it didn't get a lot of attention from US news outlets. But it was and still is being covered by several science magazines. Scientists are currently studying the effects and environmental impacts from the record-breaking amounts of sulfur dioxide released into our atmosphere over the six-month eruption.

Most Active Earthquake Area in Iceland

Brimketill lava rock pool, "Oddny's pool," located on the peninsula of Reykjanes , Iceland is the area where earthquakes are the most active and only eight miles from where the Bridge Between Two Continents is located. At the end of January 2019, within a 24-hour period, there were 149 recorded earthquakes in Iceland. We may not have heard this news not because it wasn't newsworthy to the world, but because it's a regular occurrence in Iceland.

Lava Rock

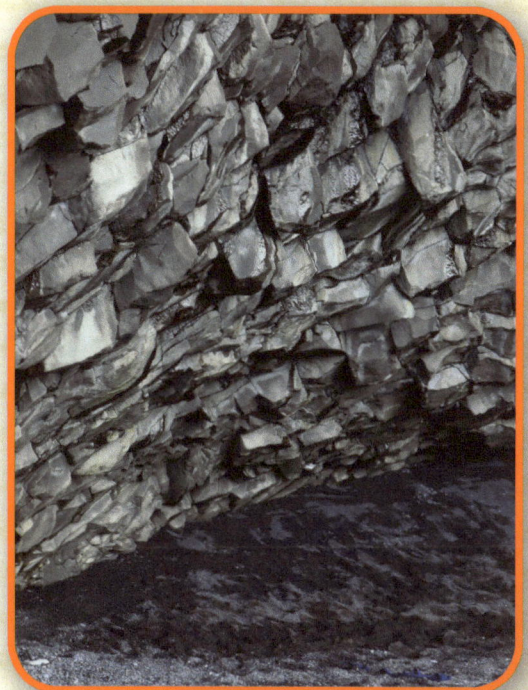

Lava rocks, or volcano rocks, are formed from magma from an erupted volcano. Eighty percent (80%) of Earth is covered in volcanic rocks. There are several uses for volcano rocks because they retain heat and water naturally. People use them in grills because they distribute heat evenly. They are used in landscapes because they retain water and heat, helping plants and landscaped yards. They are also great for aquariums because they act as a natural water filter.

Basalt Columns

Basalt is a very common volcanic rock, or lava stone, and it has a very low amount of silica. Basalt columns are formed when lava crystalizes within a lava flow. It reaches the Earth's surface at temperatures between 1100° to 1250° C (2012° to 2282° F) and then a small section of lava flow begins to cool rapidly from the inside out, which causes swelling, thus breaking off and making a fairly even column formation. The heat from the lava escapes regularly, repeating these steps, and developing into basalt columns. These columns are simply fascinating to see.

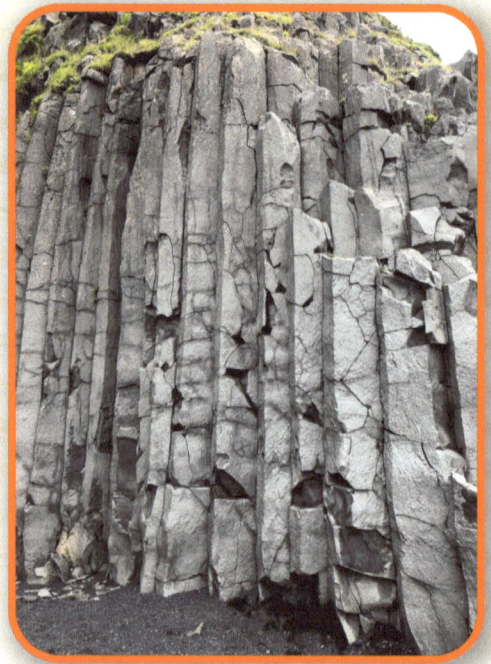

Ancient Moss

Moss isn't listed as one of the Wonders of the World, but maybe it should be. There are 606 different species of moss in Iceland. Woolly fringe-moss retains water and humidity, prevents soil erosion, and has no root system. Moss that grows as large mats on top of exposed lava rocks can grow up to a foot and a half thick, but it takes hundreds of years to achieve this size. It's sad to say, but because of this moss's sensitivity, if someone steps on it—just one step—it will die along with all the microorganisms living inside it. So, when visiting Iceland, please stay off the moss. It is very tempting to have a picture taken sitting on the moss, but is it worth the hundred years it took to grow?

Glaciers

Nine thousand years ago, the volcanoes erupted and melted all ice caps and glaciers, and with each eruption, the island formed—the land we now know as Iceland. Around 2,500 years ago, Iceland started the freezing process over again. This process entailed years of snow/freeze/rain/melt a little, snow/freeze, and again snow/freeze some more, repeated for thousands of years, the snow packing itself down, compressing into ice masses

and creating these thick glaciers known as ice sheets or ice caps.

A glacier is the layering of compressed snow formed into large, thickened ice masses. To be considered a glacier, the compressed ice mass must be thick enough to sink and move under its own weight. Small ice sheets covering up to 50,000 square kilometers (19,305 square miles) are called ice caps. Glaciers covering over 50,000 square kilometers are called ice sheets. Greenland and Antarctica are covered in these large ice sheets. Antarctica is over three miles thick in some areas. Now that's some glacier!

Glaciers are basically frozen lakes on top of mountains and volcano tops, with frozen rivers coming down the mountain valleys. Vatnajökull is 1,300 feet thick, and the tongues, or rivers, coming down from the top are wide because the decades or centuries of heavy weight from the ice pushed the ice out farther at the edges of the rivers, causing the rivers to widen. We can watch the running water at a river's edge widen in minutes, hours, days, or weeks, but a frozen river would take decades to move inches if it remained frozen.

Eleven percent (11%) of Iceland is covered by glaciers. It has 269 named glaciers, including ice caps, outlet glaciers, mountain glaciers, alpine, piedmont, and cirque glaciers, and ice streams. Many of Iceland's glaciers are ice caps that have been created on top of volcanoes .

Vatnajökull means "the river glacier" and is nicknamed "The Big One" because it covers 8% of Iceland. This is one of Europe's largest glaciers because of how thick (measured in volume) the condensed ice is.

The glacier with the longest name is Eyjafjallajökull and, ironically, it is one of the smallest glaciers in Iceland. Eyjafjallajökull is also the name of an ice-capped stratovolcano where both the volcano and the glacier bear the same nearly unpronounceable name since this ice cap covers the volcano's caldera, or sink hole, that was created by collapsed magma chambers.

The smallest glacier in Iceland was Okjökull, but due to climate change and warmer weather, it recently lost its *jökull*—glacier—status and is now simply Ok. Unfortunately, this isn't an American joke. Okjökull has been downgraded from being a glacier because rising temperatures have caused it to shrink. As a matter of fact, ice hardly exists there now, and that's not okay, because Arctic ice is important—more important than most of us have ever considered.

Why Arctic and Antarctica Ice Are Important

Like penguins, Arctic fishes, seals, and polar bears are heavily dependent on sea ice for nearly everything, as is much of their prey. Floating glaciers are rapidly disappearing, making it nearly impossible for polar bears and Arctic foxes to hunt, travel, mate, rest, even provide shelter for their newborn young. Nearly all ocean life is being affected by warming ocean temperatures, to the detriment of Earth's entire interrelated ecosystem.

While out at sea, Penguins rest and find a safe-haven on floating icebergs such as this one. (Antarctica 2020)

Glaciers and glacier rivers have been frozen for thousands of years and took thousands of years to create. Now that our world's temperatures are rising, the freeze period isn't as long and the melting period is longer, so thawing is happening much faster than people once believed. Scientists say glaciers are melting at an alarming rate and are warning us about what is to come. The glaciers' tongues are receding faster, and huge chunks of glaciers are falling into these lagoons, floating away, and melting at record speed. These melting glaciers send more water to sea and more moisture into the air. This is creating crazy storms, floods, and an alarming rise of sea levels around our planet.

The importance of Arctic sea ice is incalculable. It helps moderate global temperatures with its ability to reflect sunlight back into space. What happens in the Arctic doesn't stay in the Arctic—it affects all the rest of our planet.

Walking on Glaciers for Fun

In Iceland, glacier-walking tours are some of the most popular tourist activities, besides glacier climbing. Glacier activities can be done all year, making glacier tours one of the top tourism industries in the country. Needless to say, glaciers are slippery! When walking over the crevasses and ice ridges, you must wear your most comfortable, warmest hiking boots with special ice cleats, or traction cleats. These have special spikes to prevent people from slipping and falling down. You must also wear a helmet, because even when wearing these spikes on your shoes, you may fall! There are several ice tunnels and caves in Iceland that tour guides can walk you into, including both down and inside ice caves inside volcanoes like Vatnajökull. Yes, most glaciers in Iceland rest above active and inactive volcanoes!

Waterfalls

Waterfall is *foss* in Icelandic, so most of the waterfall names end in *foss*. There are over 10,000 waterfalls in Iceland due to the frequent rain, snowfall, and large glaciers feeding the rivers. The water is extremely cold, clear, and clean, so it's pure enough to drink without filters and tastes unbelievable.

Hundreds of waterfalls can be seen from the main road that loops around Iceland and there are also waterfalls you can only find after a long hike off the beaten path. Some are located behind or near private homes and farms. Waterfalls can be right around the corner where you least expect them.

They are also located below the landscape, such as Gullfoss and Hjalparfoss in south Iceland. Other waterfalls are hidden, like the Gorge Dweller in book three of Iceland: The Puffin Explorers Series, *Puffins Encounter Fire and Ice.* In caves and about a hundred yards away is Seljalandsfoss waterfall. This beautiful, huge waterfall drops 197 feet (160 meters) and is a popular tourist destination because it's near the main road and there is a path so you can walk behind it.

Gullfoss

Iceland's Hvita River, meaning White River, runs approximately 1.5 miles long and is 230 feet deep. Because of constant water flow, the riverbanks have an erosion rate of about 25 centimeters every year. The rocks of the riverbed were formed during an interglacial period, the Ice Age. One of this river's most famous attractions and Iceland's most famous waterfall, Gullfoss, is also known as the Golden Waterfall because on sunny days it shines a golden color. This waterfall projects a beautiful rainbow at every angle and sits inside the golden circle, a tourist driving route, making it Iceland's most visited and most popular waterfall. It is a two-tiered waterfall, cascading 105 feet and seemingly disappearing into the earth as it falls into a volcanic rock gorge. This waterfall can only be viewed from above.

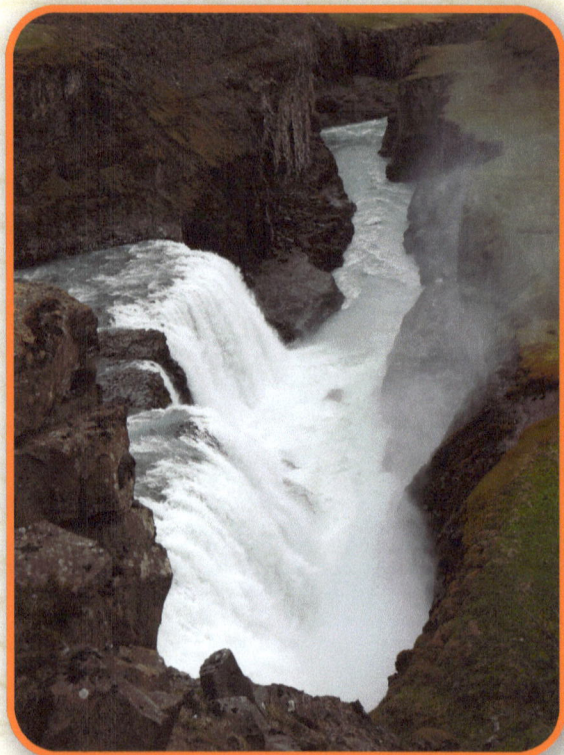

Glymur

Glymur waterfall was considered Iceland's tallest at over 198 meters, but in 2011, Morsárfoss was measured at 228 meters (748 feet) and now takes the crown for being the tallest waterfall in Iceland. This waterfall falls directly from the Esjufjöll area, which is part of the Vatnajökull icecap glacier.

Goðafoss

Goðafoss (GO-thuh-foss) means "waterfall of the gods." Its water falls 12 meters (40 feet) and is 30 meters (98.4 feet) wide. When Icelanders officially adopted Christianity in the year 1000, the statues of Norse gods were thrown into Goðafoss, symbolizing their new Christian faith and giving the waterfall its name. Goðafoss is located off the main road around Iceland, approximately a 5-hour drive from the city of Reykjavik.

Daring, fearless people enjoy kayaking off Goðafoss. I waited for several hours for some people to find the courage, but they never did. My cousin visited months later and found people going off the waterfall as if it were a regular occurrence.

Svartifoss

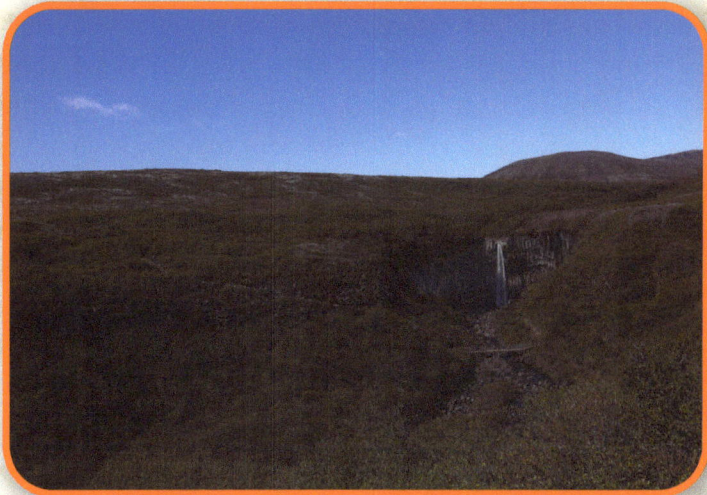

A spectacular waterfall you must hike to is Svartifoss. It is surrounded by basalt columns. It's about a 45-minute hike (1.5 kilometers), and on the way, you will see three much smaller waterfalls: the Þjófafoss (Thieves Falls), Hundafoss (Dog falls), and Magnusarfoss (the falls of Magnus).

Once you reach Svartifoss, its beauty may take your breath away. The basalt columns near the waterfall appear black due to the spray from the waterfall onto the dark gray columns making them appear darker. Rain or shine, this is a stunning waterfall.

Dettifoss

Dettifoss is located in the northeast of Iceland and is the largest waterfall in Europe (in measured volume). The average volume—water flow—discharged is 200m³/s (cubic meters per second), which translates to about 3,170,074 gallons of water per minute. That's an impressive waterfall!

Green Power

Iceland is one of the world leaders when it comes to the use of renewable energy sources, and it is the eco-friendliest nation in the world. Renewable energy is energy that can be replenished continually by nature. The country gets 100% of their heating and electricity needs from renewable energy sources —about 75% from hydropower and 25% from geothermal energy.

Hydropower is energy generated by the movement of water. When you see the prefix "hydro," you know it has something to do with water. It can be produced with water, wind, steam, an old-fashioned wooden water wheel, and even hand cranks. It can come from the movement of waterfalls, ocean waves, a rushing stream or river, or water being propelled through pipes. Hydropower is one of the oldest forms of energy, dating back to 4000 BCE. Iceland has over 10,000 waterfalls, and several hundred are large enough to tap into for energy.

Geothermal energy is heat naturally produced and stored underground. When you see "geo" you know it has something to do with the Earth, and when you see "thermal," you know it has something to do with heat.

This land of fire and ice has abundant sources of "green" electricity. Active volcanoes and their underground magma play a huge role in geothermal power. Cold rainwater seeps through the Earth's surface and reaches the magma intrusions, thus heating the water— and it doesn't just heat it, it makes it boil !

As a result, it rises back to either the Earth's crust or all the way to the surface as boiling water or steam. This process creates the hot springs, geysers, fumaroles (hot, stinky Earth farts), and volcanoes.

Iceland has geothermal power plants that produce electricity, harnessing geothermal energy and using it to provide heat and electricity to homes and businesses. But half of Iceland can just tap into these resources themselves, in their own back yards, to provide their own water, heat, and electricity that powers the artificial lighting necessary at such a northerly latitude. This is especially needed in winter, since December in Iceland can be dark virtually 24 hours a day.

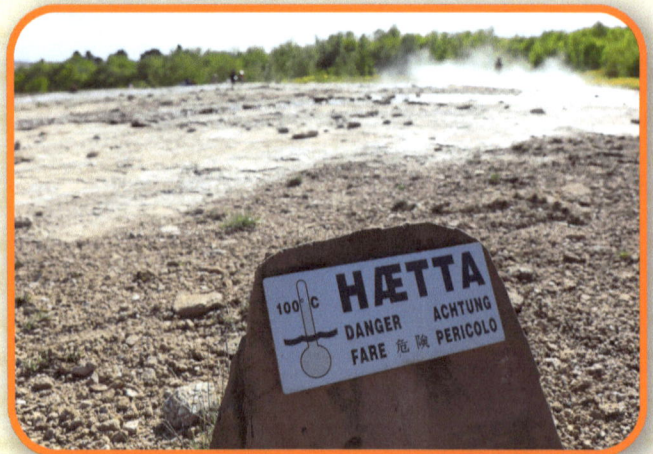

Underground reservoirs are tapped for geothermally heated water or steam. The energy is captured, or harnessed, by drilling wells into the Earth. When water travels down the cracks and reaches the hot rocks, which are heated by the molten lava underneath, the reaction creates steam. The steam rises and is then converted into geothermal heat and electricity to heat and power homes, schools, hospitals, and businesses. In most places on Earth, this energy isn't seen from above the ground, but you can actually see these openings in several forms all over Iceland.

One of Iceland's oldest and most important uses of geothermal energy is for greenhouses. Most of Iceland's greenhouses are enclosed with glass and located in the south. Geothermal steam is used to boil and disinfect the soil, warm the greenhouses, water plants, and create energy to run artificial lighting in

the winters when it's dark most of the day, making it possible to grow crops all year round.

Geysers are another source of renewable geothermal energy in Iceland. They are the results of boiling water trying to resurface and exploding up into the air, thus releasing the pressure built up under the ground. The English word for *geyser* comes from the name of the great geyser Haukadlur, first discovered in Iceland in 1294 and the oldest known geyser on the planet. Iceland's Geysir Strokkur was first reported after an earthquake in 1789. It erupts once every 6 to 10 minutes, reaching an average 60 feet high. The water chambers below boil at about 300° F, and looping chambers trap the steam. The pressure from the heat builds until it finally releases, forcing the water out of these pools and high into the sky.

Pools and hot tubs in Iceland are heated by geothermal energy. Icy roads and parking areas are cleared off using geothermal snow-melting systems. Geothermal pools are found all over Iceland for the public to use, but use caution when in isolated areas because these pools can be boiling hot—hot enough to cook food—and not look it.

Blue Lagoon: Healing Powers from a Silica Kiss

The Blue Lagoon is located in the Reykjanes peninsula where there are miles and miles of lava fields. Nestled up against the small hills are geothermal plants and some unique hotels. The Blue Lagoon is one place where you can swim/spa/bathe in this beautiful landscape.

The Blue Lagoon has water containers (big boxes full of hot water from below) sitting over the pool areas, being cooled by the cold air before releasing it into the spa area. The Blue Lagoon water temperature is kept at about 98 to 102 degrees Fahrenheit (37 to 39 degrees C).

It is said that a man who had worked at the geothermal site next door to the Blue Lagoon had a skin disease and after months of "bathing"—soaking in these waters and rubbing the silica over his skin—the disease healed! He then started researching, and years later, the Blue Lagoon was opened to the public and became known to people all over the world.

Hot Springs Bread

Hot geothermal underground springs are used to bake Iceland's famous Hot Springs Bread, Rúgbrauð . In the town of Laugarvatn, this special bread is still made today. They mix the bread, put it in a pot, and cover the pot with a plastic wrap to keep the water out. They dig a hole 12 inches deep, place the pot inside, and bury it. Twenty-four hours later, they dig up beautiful, moist bread that is ready to eat. It is not uncommon for Icelandic families to bake this bread and share it in family gatherings.

If you eat too much of Rúgbrauð, you may then call it Thunder Bread...can you guess why?

Taking Care of Our Planet

Environmental issues are often put on the back burner even by caring people because most of us focus on our immediate problems. Long-term problems in the environment become "mid-level worries" because we tend to scan for and focus on danger and problems that need to be fixed right now. For example, it is easy to bring food and clothing to someone who lost their house in a fire. We can see what to do, and it is easy to do it. It is not so easy to fix the environment's problems, which were created over time and will take a long time to fix. But we still need to think about it and do things every day that will help. The water, the air, and all Earth's creatures are depending on us. There is no planet B .

TIDBITS, STATISTICS, AND TRIVIA

In July, Iceland has a couple hours of soft, lingering twilight. They call this the midnight sun. So basically, it is light 24/7!

Midwinter has only 4 to 5 hours of effective light and super cool northern lights that are also known as the Aurora Borealis.

By December, Iceland has almost 24 hours of darkness with some twilight at midday.

Icelandic life is conditioned by landscapes, ocean currents, and weather.

Iceland's average temperatures :

Summer
     Air: 10° to 25° Celsius (55° to 77° Fahrenheit)
     Sea: 8° to 10° Celsius (46.4° to 50° Fahrenheit)

Winter
     Air: 0° to 30° Celsius (22° below
     zero to 34° Fahrenheit)
     Sea: below 8° C (46.4° Fahrenheit)

This is why there are no mosquitoes in Iceland!

During the month of January, the temperature in Reykjavik is about the same as it is in New York to Maine in the United States.

- was formed about 25 million years ago, and is one of the youngest landmasses on the planet
- was one of the last places on Earth to be settled by humans
- was carved by glaciers and shaped by magma and earthquakes
- is the eco-friendliest country in the world (renewable energy sources)
- is the oldest democracy in the world
- Greenland was colonized by people from Iceland in 986
- At the end of the eighteenth century, 20–25% of Iceland's population emigrated to the US and Canada because of a food shortage.
- became independent (but still in union with Denmark as the Kingdom of Iceland) in 1918
- founded the fully independent Republic of Iceland in 1944
- is the eighteenth largest island in the world
- has no railway system of any kind
- has no McDonald's
- has very little crime
- has higher Coca-Cola consumption per capita than any other country
- Beer was illegal until 1989.
- Off-roading; owning a snake, lizard, or turtle; and using hand-held devices while driving are also illegal.

The Christianization of Iceland

Icelanders officially embraced Christianity around the year 1000, after a *lawspeake* named Ljósvetningagoði made his monumental decision to convert. He demonstrated his new faith by throwing the idols of the old Norse gods into the waterfall Goðafoss, located in the north region of Iceland. But the mythology of the Norse gods and their ancestral heritage continues to be felt in many aspects of Icelandic culture.

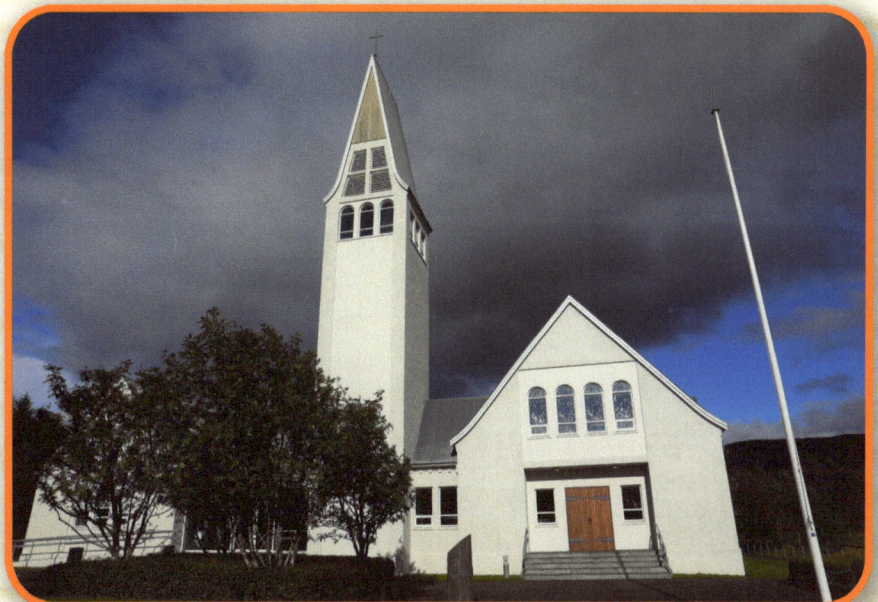

Compare the Numbers!

Size and population compared to US states:

Size
- Iceland is 39,000 square miles in area.
- Kentucky is 40,409 square miles.
- Ohio is 44,825 square miles.

Population (Estimated as of December 2018)
- Iceland – 339,747
- Kentucky – 4,472,265
- Ohio – 11,689,442

Tourism

In 2018, the estimated number of tourists visiting Iceland was almost seven times Iceland's total population at 2,315,925. That's a lot of people checking out puffins! And waterfalls, glaciers, volcanoes, horses, geysers, the northern lights, ice caves, lava caves, more waterfalls...

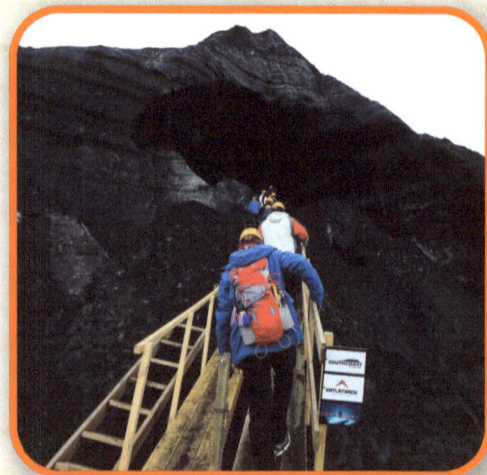

Imports, Exports, Industries

A few top industries:
- tourism
- fish processing
- aluminum smelting
- geothermal power
- hydropower
- medical and pharmaceutical products

A few examples of the $7.53 billion in imports in 2017:
- cars – 8.4%
- aluminum oxide – 8.1%
- planes, helicopters, and/or spacecraft – 7.1%
- fishing supplies – 2.7%
- delivery trucks – 1.4%
- carbon-based electronics – 4.1%
- broadcasting equipment – 1.6%
- computers – 1.45%

A few examples of the $5.63 billion in exports in 2017:
- aluminum – 36%
- fish fillets – 17%
- other fish – 13.2%
- animal meal and pellets – 3.3%
- Other:
  - fishing ships
  - passenger and cargo ships
  - scales
  - orthopedic appliances
  - medical instruments
  - other

Other Books About Puffins:

- *Arctica Puffin* by R.J. Maddocks (ISBN 9781522742135)
- *Atlantic Puffin: Little Brother of the North* by Kristin Bieber Domm, Illustrations by Jeffrey C. Domm, Nimbus Publishing (ISBN 9781551095189)
- *Discover Puffins* by Discover Books Series (ISBN 9781500228323)
- *Project Puffin: How We Brought Puffins Back to Egg Rock* by Stephen W. Kress, as told to Pete Slamansohn, A National Audubon Society Book (ISBN 9780884481713)
- *Puffins* by Drew Buckley, GRAFFEG Pocket Books (ISBN 9781909823105)
- *Puffins for Kids* by Rachel Smith, Mendon Cottage Books, JD-Biz Publishing (ISBN 9781516852543)
- *Puffling Patrol* by Ted and Betsy Lewin, Lee & Low Books Inc (ISBN 9781620141878)
- *The Puffins Are Back* by Gail Gibbons Holiday House (ISBN 9780823441631)
- *Puffins (Wildlife Monographs)* by Heather Angel (ISBN 9781901268195)

Websites

NationalGeographic.com (news, Iceland, and tracking puffins)
kids.nationalgeographic.com (puffins, puffins vs. penguins)
The Royal Society for the Protection of Birds at RSPB.org.uk
Audubon.org
birds.com
puffinpalooza.com
worldwildlife.org
https://oceanconservancy.org/blog/2018/10/24/5-feathery-facts-puffins/

Tracking Puffins:
https://www.audubon.org/news/puffin-mystery-solved-first-tracking-evidence
https://ebird.org/map/atlpuf
https://phys.org/news/2017-04-puffins-partner-migration-chicks.html
https://phys.org/news/2016-02-scientists-puffins-winter-jersey.html

SIGNS NOT TO IGNORE:

Please Stay Off the MOSS!

Cliff edge is unsecure

HÆTTA
DANGER ACHTUNG
FARE PEROCOLO

100° C

Vinsamlegast fylgið merktum gönguleiðum
um háhitasvæðið við Gunnuhver.
This is a high temperature geothermal area.
Please follow the marked pathway at all times.
Questa e una zona geotermica ad alta temperatura.
Si prega di sempre seguire il percorso segnato.

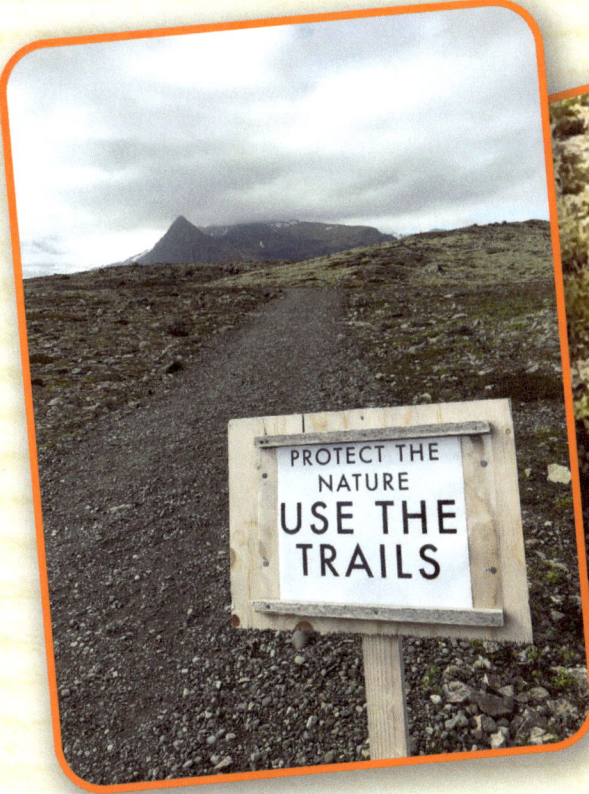

PROTECT THE
NATURE
USE THE
TRAILS

HÆTTA
DANGER ACHTUNG
FARE 危险 PERICOLO

100° C

ABOUT THE AUTHOR

RA Anderson is a wanderer who has lived all over, from California to Belize, and currently, home is a town called Rome, in Georgia that is! She grew up on horseback and sailboats—"the most amazing way to grow up!"

A lifelong passion for creative writing and photography became her life. Her award-winning photographs have been featured in table books, magazines, and front-page news, and her writing has been published in magazines, poetry books, and children's books.

Three boys—her heart and soul—call her Mom. She and her husband—"my strength and passion"—are recent empty-nesters, leaving them more time to travel.

"My life is full, colorful, and exhausting, and I wouldn't trade it for anything. However, people seem to think my most impressive accomplishment is that I know how to work the manual settings on a DSLR camera!"

OTHER BOOKS BY RA ANDERSON

If Pets Could Talk: Dogs

If Pets Could Talk: A Service Dog

If Pets Could Talk: Cats

If Pets Could Talk: Farm Animals

Girl Sailing Aboard the Western Star

Puffins Off the Beaten Path (Iceland: The Puffin Explorers Series Book 1)

Puffins Take Flight (Iceland: The Puffin Explorers Series Book 2)

Puffins Encounter Fire and Ice (Iceland: The Puffin Explorers Series Book 3)

www.ingramcontent.com/pod-product-compliance
Lightning Source LLC
Chambersburg PA
CBHW041240020426
42333CB00002B/28